SCIENTIFIC AMERICAN EDUCATIONAL PUBLISHING

WHAT A REACTION

10 FUN CHEMISTRY EXPERIMENTS

BRING SCIENCE HOME

Published in 2025 by Scientific American Educational Publishing in association with The Rosen Publishing Group.

Contains material from Scientific American, a division of Springer Nature America, Inc., reprinted by permission as well as original material from The Rosen Publishing Group.

First Edition

Editor: Kristen Rajczak Nelson
Designer: Rachel Rising

All illustrations by Continuum Content Solutions

Activity on page 5 by Megan Arnett and Science Buddies (August 25, 2016); page 11 by Science Buddies (January 5, 2012); page 17 by Science Buddies (June 7, 2012); page 23 by Science Buddies (May 2, 2013); page 29 by Science Buddies (June 25, 2015); page 33 by Science Buddies (December 31, 2015); page 39 by Science Buddies (April 28, 2016); page 44 by Svenja Lohner and Science Buddies (January 17, 2019); page 49 by Science Buddies and Sabine De Brabandere (July 18, 2019); page 55 by Science Buddies (June 14, 2012).

Photo Credits: pp.3, 4, 11, 14, 17, 20, 23, 26, 39, 42, 43, 44, 47, 55, 59 cve iv/Shutterstock.com; pp. 5, 11, 17, 23, 29, 33, 39, 44, 49, 55 Anna Frajtova/Shutterstock.com.

Library of Congress Cataloging-in-Publication Data

Names: Scientific American, inc., editor. | Scientific American Educational Publishing.
Title: What a reaction : 10 fun chemistry experiments / Scientific American Editors.
Other titles: 10 fun chemistry experiments
Description: Buffalo, NY : Scientific American Educational Publishing, an imprint of Rosen Publishing, [2025] | Series: Bring science home | Includes index.
Identifiers: LCCN 2024005868 (print) | LCCN 2024005869 (ebook) | ISBN 9781725350731 (library binding) | ISBN 9781725350724 (paperback) | ISBN 9781725350748 (ebook)
Subjects: LCSH: Chemistry--Experiments--Juvenile literature. | Science projects--Juvenile literature.
Classification: LCC QD38 .W43 2025 (print) | LCC QD38 (ebook) | DDC 540.78--dc23/eng/20240223
LC record available at https://lccn.loc.gov/2024005868
LC ebook record available at https://lccn.loc.gov/2024005869

Manufactured in the United States of America

CPSIA Compliance Information: Batch #CSSA25. For further information contact Rosen Publishing, New York, New York at 1-800-237-9932.

CONTENTS

⚛ THESE ACTIVITIES INCLUDE SCIENCE FAIR PROJECT IDEAS.

INTRODUCTION

From exciting explosions to creating crystals and snow to a taste test with edible paper, science that involves chemistry is sure to be fun. Using common household items and the instructions you will find in this book, you can learn more about the world around you by conducting some chemistry experiments, either by yourself, or with the help of a friend!

Projects marked with include a section called Science Fair Project Ideas. These ideas can help you develop your own original science fair project. Science fair judges tend to reward creative thought and imagination, and it helps if you are really interested in your project. You will also need to follow the scientific method. See page 61 for more information about that.

Diapers

What Keeps Babies and Astronauts from Springing a Leak?

HOW DOES CHEMISTRY HELP SEAL AWAY WATER IN DIAPERS AND OTHER ABSORPTIVE PRODUCTS? TRY THIS GOOEY SCIENCE ACTIVITY AND FIND OUT!

PROJECT TIME

25 minutes

KEY CONCEPTS

Chemistry
Materials science
Polymers
Volume
Absorption

When we think about diapers we usually think about babies. But did you know that astronauts also have to wear diapers sometimes? Astronaut "diapers" are called maximum absorbency garments (MAGs), and astronauts wear them when they have to stay sealed in their spacesuits for long periods of time, such as during spacewalks or when their spacecraft leave orbit to reenter the atmosphere to return home.

For babies and astronauts, the most important thing for a diaper is to prevent leaks. But did you ever wonder how a disposable diaper actually works? What's inside the diaper that allows it to absorb all that wet stuff without making a mess? In this activity, we're going to explore the substance in diapers that allows them to stay leak-free—in a crib and in outer space!

BACKGROUND

Whether it is made for a baby or an astronaut, the major disposable diaper brands all contain a powdery chemical absorbent called sodium polyacrylate, which can absorb over 300 times its weight in water! That's about equivalent to a normal adult being able to suddenly absorb the weight of a blue whale, about 40,000 pounds (18,144 kg)!

Being able to absorb a lot of liquid is what makes sodium polyacrylate important in diapers. The chemical absorbs the liquid in the diaper, pulling it away from the skin. The diapers that astronauts wear can absorb 1/2 gallon (2 l) of water—the same amount of liquid that's in a big soda bottle at the grocery store! And thank goodness that they can absorb this much, because nobody wants to have a wet diaper while they're fixing a spaceship!

In this activity, you're going to separate sodium polyacrylate from diapers and test its absorbency (how much liquid it can hold) for yourself. One way you will measure this is by measuring volume, which means the amount of space that a substance occupies. In this case, you'll be measuring the volume when you mix different amounts of sodium polyacrylate with water.

MATERIALS

- Adult helper
- Eight clear plastic cups (A glass will also work, as long as it is clear.)
- Two to three disposable diapers (The number needed will vary slightly depending on size and brand.)
- Food coloring
- Permanent marker
- Scissors
- One clean, gallon-size, zip-top bag
- Measuring spoon
- Measuring cup
- A spoon or something else to use as a stirrer
- A workspace that can get wet and/or newspaper to cover it
- A teaspoon of table salt (optional)

PREPARATION

- If desired, spread a few sheets of newspaper over your workspace. The chemical you'll be working with is perfectly safe, but this will help with cleanup.
- Use the permanent marker to label four of the cups. (If you're using glass cups, write the labels on a piece of tape and stick the tape to the cup.) Label the cups as follows: cup 1—1/8 teaspoon, cup 2—1/4 teaspoon, cup 3—1/2 teaspoon, cup 4—1 teaspoon.
- Place two to three drops of food coloring in the four unlabeled cups. Use the measuring cup to pour 1/3 cup (79 ml) of water into each cup. Use a spoon or stirrer to mix the coloring and water.
- Open a diaper on your workspace. Have an adult help you use the scissors to cut through the inside lining of the diaper and remove the powdery, cottonlike stuffing material inside. Place the stuffing in your zip-top bag.
- Check the diaper for any leftover powder—either scoop this up with your hands (or use your spoon) or you can carefully pick up the diaper and shake the remaining powder into the zip-top bag.
- Set the diaper aside. With all the stuffing inside the bag, carefully seal it and shake the bag for one to two minutes. You should see more powder collecting at the bottom of the bag as you shake it loose from the stuffing.
- Carefully remove the stuffing from the zip-top bag. (It is okay if there is a little left inside.)
- If necessary, repeat this process with one to two more diapers until you have collected at least 2 teaspoons of the powder in your bag.

PROCEDURE

- Use your measuring spoon to transfer the appropriate amount of powder from the bag to the corresponding cup. Start by putting 1/8 teaspoon of powder into the cup labeled "1/8 teaspoon," then 1/4 teaspoon of the powder into the cup labeled "1/4 teaspoon," etcetera. Continue until you've put the correct amount of powder into each labeled cup. Notice the look and texture of the powder. *How much space does it take up in the cup?*

- With help from an adult, carefully pour 1/3 cup (79 ml) of the colored water you measured into the cup labeled "1/8 teaspoon." Use your spoon to stir the water and powder. Stir the liquid for approximately one minute. *What do you notice about the water as it mixes with the powder? Is the consistency changing as you stir? Can you still see the powder?*

- Repeat this process, pouring 1/3 cup (79 ml) of the colored water you measured into the cup labeled "1/4 teaspoon." Use your spoon to stir the water and powder for approximately one minute. *What do you notice about the water as it mixes with the powder? Is the mixture's appearance changing as you stir? What about the consistency of the mixture? Is it different from the mixture in the 1/8-teaspoon cup?*

- Repeat this process, pouring 1/3 cup (79 ml) of the colored water you measured into the cup labeled "1/2 teaspoon." Stir the water and powder for approximately one minute. *What do you notice about the water as it mixes with the powder? Is the mixture's appearance changing as you stir? What about the consistency of the mixture? Is it different from the mixture in the 1/4-teaspoon cup?*

- Repeat this process, pouring 1/3 cup (79 ml) of the colored water you measured into the cup labeled "1 teaspoon." Stir the water and powder for approximately one minute. *What do you notice about the water as it mixes with the powder? Is the mixture's appearance changing as you stir? What about the consistency of the mixture? Is it different from the mixture in the 1/2-teaspoon cup?*

- Line up all four of the labeled cups. You can observe the volumes of the mixtures in each by noticing the height of the liquid in the cup labeled "1/8 teaspoon." Compare this to the heights of the mixtures in the other cups. *Which cup has the largest volume? Which has the lowest? What other differences do you notice about the mixtures in the different cups? Do they look different? Do they feel different?*

EXTRA

Using your measuring spoon, add 1 teaspoon of salt to the mixture in the cup labeled "1 teaspoon." Stir the mixture for about one minute. *What do you notice happening to the mixture as you stir? Is it changing? What is different about it after adding the salt?*

EXTRA

Try increasing the amount of powder up to 1 tablespoon, always adding 1/3 cup (79 ml) of water. Notice how the consistency changes as you increase the amount of powder.

EXTRA

If you have an extra diaper, try soaking it in water for a minute, then set it somewhere to dry. Observe whether the diaper changes shape or texture after being soaked in the water.

OBSERVATIONS AND RESULTS

As the amount of powder increased in each cup, you may have noticed that the volume of the mixture increased. In addition, you should have noticed that the texture of the powder changed. When there was enough powder to absorb all of the water, the powder went from tiny crystals to larger gel spheres, almost the consistency of round pieces of rice. This is the expected result. Sodium polyacrylate is a polymer, which means it is a large chain of molecules made up of many smaller units known as monomers. Superabsorbent polymers expand when they come into contact with liquids like water, because water is drawn into and held by the molecules of the polymer. Therefore the sodium polyacrylate acts like an extremely powerful sponge!

If you tried adding salt to the mixture, you might have noticed that the texture and volume of the gel mixture changed back to a liquid. This is because salt breaks the attraction between water and polymer chains. The added salt breaks into negative and positive ions, which are also attracted to the water molecules. Each salt ion attracts several water molecules to it, thereby pulling them away from the polymer. Eventually enough water is pulled away so that the mixture goes from gel back to liquid.

CLEANUP

Throw out any garbage. Wash materials with soapy water and put them away once they are dry.

Playing with Polymers

CREATE YOUR OWN POLYMER ... AND HAVE FUN PLAYING WITH IT AFTERWARDS!

PROJECT TIME

20 minutes

KEY CONCEPTS

Chemistry
Polymers
Ratios
Mixtures

Have you ever wondered how fun toys like Silly Putty, Gak, and Slime are made? It's the properties of polymers, certain kinds of large molecules, that make these products bouncy, slimy, stretchy, breakable, hard, soft, sticky, moldable—and just plain fun to play with.

Polymers are found in a variety of materials that have a broad range of properties. Materials made from polymers can be found in nature, such as amber and natural rubber, or generated synthetically, such as nylon, silicone, and all plastics. The unique physical and chemical properties of polymeric materials can change depending on the amount of each different ingredient used to make them. How will changing the ratio of ingredients affect how a polymeric material feels and behaves?

BACKGROUND

You can make a polymer-based material similar to Silly Putty at home by mixing together water, borax, and Elmer's Glue. Elmer's is made up of polyvinyl acetate, which is a synthetic polymer. A polymer is a long molecule that is mostly made up of many similar repeating units. In the case of polyvinyl acetate, each repeating unit contains an acetate group. Borax, which is a white powder made up of sodium tetraborate, can react with this acetate group. Specifically, one molecule of borax can react with acetate groups on two different polyvinyl acetate molecules (such as those from the glue), creating a bond between the two polyvinyl acetate molecules.

Borax cross-links the polyvinyl acetate molecules. The more cross-linked molecules, the larger the polymeric material that is made from the reaction. Additionally, as more cross-links are made, the more the polymeric material becomes less liquidlike and gains solidity.

MATERIALS

- Three zip-top bags
- Water
- Elmer's white school glue
- Borax (also called 20 Mule Team Borax household cleaner)
- Measuring cups
- Measuring tablespoons and teaspoons
- Two glass jars with lids
- Permanent marker
- Gloves, latex, or similar style exam glove (optional)
- Goggles or other eye protection

PREPARATION

- Children should wear goggles or other protective eyewear, and adults should supervise and use caution when handling borax because it can irritate eyes.
- In one glass jar, add 3 tablespoons (44 ml) of water and 3 tablespoons (44 ml) of Elmer's Glue. Tightly secure the jar lid and shake it until the glue is fully diluted and no gooey clumps remain. Label this jar "Solution #1: 50 Percent Glue."

- In the second glass jar, add 1/2 cup (118 ml) of warm water and 1 teaspoon of borax. Again, tightly secure the lid and shake it until no particles of borax remain and the solution is clear. Label this jar "Solution #2: 4 Percent Borax."

- The polymer material made in this activity can be sticky, so it should be kept off of clothing, wood, and other rough surfaces that can be hard to clean. Also, the 50 percent glue solution and the glue-borax mixtures should not be poured down a drain because they can form clogs. Dispose of them in plastic bags.

PROCEDURE

- Take the three zip-top bags and label one "A," the next "B" and the third "C."

- From the jar containing Solution #1 (50 percent glue), add 1 tablespoon (15 ml) to bag A, 2 tablespoons (30 ml) to bag B, and 3 tablespoons (45 ml) plus 1 teaspoon (5 ml) to bag C. After adding Solution #1 to each bag, carefully set the bags so they will not spill over.

- From the Solution #2 (4 percent borax) jar, add 3 tablespoons (45 ml) to bag A, 2 tablespoons (30 ml) to bag B, and 2 teaspoons (10 ml) to bag C. Seal the bags closed.

- *What are the ratios of the two solutions that were added to each bag? Which bag contains the most glue, and which contains the least?*

- With the bags sealed closed, use your fingers to squish the mixtures around in each bag, mixing together the ingredients within the bag. Some bags may require more mixing than others. *What do you observe happening within each bag? How are the resultant polymeric materials in each bag different from each other?*

- After a mixture has formed a sticky glob, you can take it out of the bag. *What physical properties can you describe for each material you've made? Are some materials runny, slimy, sticky, hard, soft, bouncy, etcetera?*
- After investigating them, if you want to save your polymer products for later, put them back in their zip-top bags, seal them, and store them in the refrigerator.

SCIENCE FAIR IDEA

Are there other ways to change the recipe in order to change the physical properties of the polymeric product? Try changing the percentages of glue in Solution #1 or of the borax in Solution #2 to see how that changes your product. *Can you optimize the recipe in a new and different way to obtain different types of products?*

EXTRA

The protein gelatin found in Jell-O is also a polymer. *What experiments could you conduct to explore the physical properties of other polymers, such as gelatin?*

OBSERVATIONS AND RESULTS

Did each bag have some solid product (a polymer material) take form inside of it after mixing? Did bag C contain the solidest polymeric material, and did bag A contain the most liquid-like one?

Elmer's Glue contains polyvinyl acetate molecules, which are long polymer molecules that are tangled with each other. This is what makes glue viscous, or thick and sticky. When borax (sodium tetraborate) is added to polyvinyl acetate and cross-links the latter's molecules to each other, the glue solution becomes more viscous.

As borax cross-links more and more of the glue molecules together and they become more viscous, an increasingly larger and more solid polymeric material is made from the reaction. The bag with the least amount of glue, bag A, should have been the most liquid-like, whereas the bag with the largest amount of glue, bag C, should have been the most solid. Store-bought Silly Putty and Slime are not made using polyvinyl acetate, but rather from organosiloxane polymers or polyvinyl alcohol to increase their durability.

CLEANUP

The 50 percent glue solution and the glue and borax mixtures should not be poured down drains as they can form clogs. Dispose of them in plastic bags that you can throw in the garbage. Wash your hands with soap and warm water as well as other materials you'd like to use again.

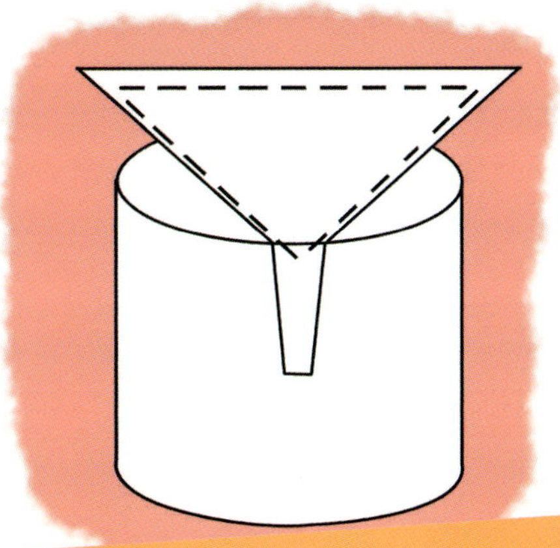

Salty Science

How to Separate Soluble Solutions

LEARN HOW TO SEPARATE SALT FROM WATER IN A SIMPLE EXPERIMENT THAT FOCUSES ON SOLUBILITY.

Have you ever mixed sand and salt together and wondered how you could separate them again? If you had to separate them, would you have nightmares of tiny tweezers, a magnifying glass, and hours spent picking grains of salt and sand apart? Don't be afraid, there is another way! Using the differences in solubility between salt and sand, you can find the simple "solution" to this problem.

PROJECT TIME

2 hours

KEY CONCEPTS

Chemistry
Solubility
Water
Temperature

BACKGROUND

Chemistry for the most part is the study of matter and how it behaves and interacts with other kinds of matter. Everything around us is made of matter. One important property that matter has is solubility. We think about this when we dissolve something in water. If a chemical is soluble in water, then when you add it to water it will dissolve or disappear. If it is not soluble, then the chemical will not dissolve and you can see it, either suspended in the water or at the bottom of the container.

When a chemical is dissolved in a liquid, such as water, it creates a solution. In a solution, the liquid is the solvent, and the soluble chemical that is added to and dissolves in the liquid is the solute. Even though a solvent dissolves a solute, the latter blends in but is still there. If you evaporated all of the liquid from the solution, you would be left with the dry solute again. In fact, this is how salt is processed in giant salt flats where seawater is slowly evaporated, leaving behind huge amounts of sea salt.

MATERIALS

- Strainer (optional)
- Salt
- Sand
- A napkin
- Magnifying glass
- Two glass jars with lids
- Measuring cup
- Water
- Teakettle or pot
- Two spoons
- Stove and oven
- Funnel
- Coffee filter
- Spoon
- Oven mitts

PREPARATION

- If the sand has a lot of debris in it, use a strainer to strain out the large debris and purify the sand.

- Place some salt and sand separately on a napkin and, using the magnifying glass, closely examine the salt and sand. *What do you notice? How does the size, shape, and color of the grains of sand compare with the grains of salt?*

- Be careful when using the stove and oven, and when handling the boiling water. An adult should help you with these steps.

PROCEDURE

- In a glass jar, add 1/4 cup of salt and 1/4 cup of sand. Put the lid on the jar and shake until the salt and sand are completely mixed together.

- Using the magnifying lens, closely examine the mixture. *What do you notice? Can you still see the individual grains of salt and sand?*

- Fill the teakettle or pot with at least 1 cup (237 ml) of water. Heat the water on the stove until it is boiling. Be careful when using the stove and handling the boiling water. An adult should help you with this.

- Very carefully pour 1/2 cup (118 ml) of boiling water into the jar and stir the mixture with a spoon. Be careful when handling the boiling water, which will make the jar become very hot! (Caution: You should pour the water very slowly, so the glass jar does not shatter from a rapid change in temperature.)

- Using the magnifying lens, closely examine the solution. *What do you notice? Can you still see the individual grains of salt and sand?*

- Place the coffee filter in the funnel and place the funnel on top of the second glass jar. Slowly pour the entire solution over the filter. As the solution seeps through the filter, let it collect in the second jar.

- *Looking at the coffee filter, what do you see?* Carefully scrape off any particles from the coffee filter with a spoon and place them in the first jar.

- Turn the oven on to 325°F (163°C). Be careful when using the oven and ask an adult to help you with this. Place both jars on a cookie sheet and bake them in the oven until all of the water has evaporated from them. This may take over an hour. *When you check on the jars to see how much water has evaporated, what do you notice?*
- Turn off the oven and let it cool down. (When glass changes temperature very quickly, it can shatter.) Then, using oven mitts, carefully remove the jars and allow them to cool to room temperature. They will probably still be very hot!
- After the jars are cool, closely examine their contents using the magnifying glass. *What do you notice? Can you still see the individual grains of salt and sand? Are they mixed together or separated?*

EXTRA

At the end of this activity, you can carefully use a measuring cup to measure the amount of salt and sand you ended up with. *Do these amounts match the amounts you started with? Why do you think this happened?*

EXTRA

Many different chemicals have different degrees of solubility. By adding different amounts of salt, sugar, or baking soda to water you can see how soluble each chemical is. Just add each chemical, 1 teaspoon at a time, to a half-full glass of water until you notice that it no longer dissolves when you stir it. *Which chemical is the most soluble (dissolves the most into the same amount of water); which chemical is the least?*

SCIENCE FAIR IDEA

How might temperature affect the solubility of a chemical? Try dissolving the same amount of sugar in hot water, room-temperature water, and ice-cold water, using the same amount of water each time. *What happens? Can you think of other variables that might affect solubility?*

OBSERVATIONS AND RESULTS

When you added the boiling water, did the salt dissolve or disappear? After drying in the oven, did the salt appear in the second jar and mostly the sand was in the first jar?

Salt is soluble in water whereas sand is insoluble (not dissolvable) in water. If sand were soluble in water, we would not have beautiful sandy beaches! Because of this, when the boiling water was added to the mixture of salt and sand, the salt should have dissolved, or disappeared, whereas the sand stayed visible, creating a dark brown solution with possibly some sand particles stuck on the walls of the jar. Temperature can affect the solubility of a chemical, and in the case of salt in water the hot temperature of the boiling water improved the salt's ability to dissolve in it. The dissolved salt should have easily made its way through the coffee filter and into the second jar whereas the muddy, undissolved sand particles became stuck on the coffee filter (possibly mixed with some salt as well). After the liquid in the two jars was evaporated in the oven, the salt in the second jar should have become apparent again, mostly as a crusty white substance along the sides and the bottom of the second jar.

CLEANUP

Wash materials with soapy water and return everything to where you found it.

Solubility Science

How to Grow the Best Crystals

DID YOU KNOW THAT YOU CAN GROW YOUR OWN CRYSTALS AT HOME? TRY THIS EXPERIMENT TO SEE HOW!

Have you ever wondered how crystals are made? Crystals come in all different shapes and sizes. The purest and cleanest crystals, however, are usually also the largest. In this activity, you'll compare the size and shape of crystals grown in different temperatures. With just water and borax, a household cleaning product, you can discover the method for growing large, pure crystals!

PROJECT TIME

20 minutes, plus 5 or more hours for crystals to grow

KEY CONCEPTS

Chemistry
Solubility
Saturation
Crystals
Purification

BACKGROUND

Chemical reactions are constantly happening all around you—and inside of you. For instance, a chemical reaction can turn metal into reddish-brown rust (the iron in the metal is reacting with the oxygen in the air or water, and the product is what we recognize as rust). Chemists perform chemical reactions to change one chemical compound into another. Sometimes when multiple products are formed, the chemist may want to separate one compound from the others. One way this can be done is using a process called recrystallization, where a solution of compounds can be dissolved in hot water and then cooled. As it cools, one substance crystalizes (becomes crystals), and can be removed from the rest of the liquid, which holds the other compound.

Why do crystals appear as the mixture cools? It has to do with solubility, or the largest amount of something that can be dissolved in something else, such as dissolving the powdered cleaning product, borax, in water. The solubility of most solids increases with temperature. In other words, more borax may be dissolved in hot water than cold water. So if a hot, saturated mixture is cooled, there's more borax than can be contained by the colder water, and so borax may fall out of the mixture, forming crystals.

MATERIALS

- Large bowl
- Ice cubes
- Water
- String
- Scissors
- Two pencils
- Two identical jars or large drinking glasses
- Cooking pot
- Borax, also called 20 Mule Team household cleaner. It can be found in the cleaning aisle of many grocery stores. (Use caution when handling cleaners—they can harm skin and eyes and should not be inhaled.)
- Measuring tablespoon
- Plastic wrap

PREPARATION

- Fill the large bowl half full of ice cubes and then add water until the bowl is about three-quarters full.

- Cut two pieces of string (they should be at least as long as the height of the jars or large drinking glasses). Tie the end of one string around each of the two pencils. Adjust the strings' lengths so that when the pencil is laid across the top of one of the jars or large drinking glasses, the end of the string hangs down to just above the bottom of the jar. Make the strings equal length.
- Borax is harmful if swallowed or inhaled or if it comes into contact with one's eyes, and on rare occasion touching it can result in rashes. Caution and adult supervision are advised when handling it.

PROCEDURE

- Fill a cooking pot with enough water to fill both jars nearly full. Then bring that water to a boil on the stove. Once the water is boiling, turn the burner off so that the water stops boiling. (Because borax is harmful if inhaled or if it comes into contact with eyes, it is advised to not dissolve it with boiling water.)
- Add 1 tablespoon of borax to the water and stir until it dissolves. Continue to add 1 tablespoon at a time until no more dissolves. You will probably need about 3 tablespoons of borax for each cup of water. *How does the saturated solution look?*
- Carefully pour equal amounts of the saturated borax solution into the two jars. Each jar should be about three-fourths full.
- Lay a pencil across the top of each jar so that the string hangs down into the saturated solution.
- Cover the top of the jars with plastic wrap.
- Leave one jar undisturbed on a countertop or table at room temperature. Place the second jar in the bowl full of ice that you prepared. If needed, adjust the water level in the bowl so that the water reaches at least three-fourths the way up the jar, but is not so high that it goes into the jar.

- Do not disturb the jars for at least five hours. Check the bowl of ice regularly and add ice if it has melted.
- Check on the jars about once an hour to see how the crystals are forming. It may be difficult to observe the jar in the bowl. Try looking at the string through the plastic wrap cover. *Do you see crystals forming on the side of one of the jars? Do crystals form in one jar before the other?*
- After at least five hours, carefully remove the pencils and observe the crystals on the strings. *How do the size, shape, and number of crystals on each string compare with one another? Why do you think this is?*

SCIENCE FAIR IDEA

In this activity, you examined borax crystal formation at two different temperatures, but you could try other temperatures as well; one way is to put one of the jars in the refrigerator. *How does allowing the borax mixture to cool at a different temperature affect crystal formation?*

SCIENCE FAIR IDEA

Try making crystals out of other materials, such as sugar or salt. *How well do crystals form using other mixtures with water?*

SCIENCE FAIR IDEA

You did this activity for at least five hours. *How do your results change if you grow your crystals for a longer period of time?* Make sure to keep adding ice cubes to the water bath to keep it cool throughout the activity.

OBSERVATIONS AND RESULTS

Did smaller, more abundant crystals form in the jar and on the string in the bowl of ice water, whereas larger, fewer, better-shaped crystals form in the jar at room temperature?

As the hot, saturated mixture of borax and water cooled, there was more borax than could be contained by the colder water, and so this borax fell out of the mixture and formed crystals. A crystal is made of molecules of a product that have come together in a specific repeated pattern. When the molecules of the crystal come together, other products that are often considered impurities, or the unwanted products of the chemical reaction, do not fit well into the structure, much like the wrong piece of a puzzle does not fit. If the crystals form slowly enough, the impurities will be rejected because they do not fit correctly, and instead will remain in the water. This is why the crystals in the room-temperature jar should have been larger and more cube-like. But if a solution is cooled too quickly, there isn't time to expel the impurities and instead they become trapped within the crystal structure and the pattern is disturbed. Consequently, the crystals in the bowl of ice water should have formed more quickly and in greater numbers but were smaller and less cubelike in shape because they had more impurities.

CLEANUP

Throw your mixtures in the garbage, and then wash materials with soapy water and return them to where you found them.

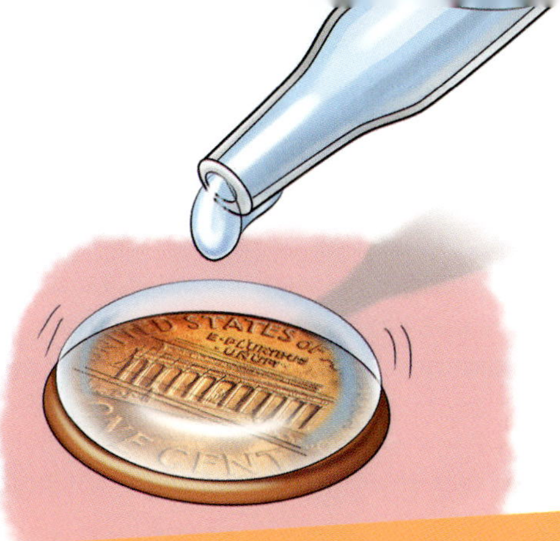

Measure Surface Tension with a Penny

LEARN ABOUT THE SECRETS OF SOAP IN THIS SURPRISING SURFACE TENSION-TESTING ACTIVITY!

Have you ever noticed on a rainy day how water forms droplets on a window? Why does it do that instead of spreading out evenly over the whole surface? You might not guess it, but this property of water is also related to washing dishes and doing the laundry. How? It all has to do with something called surface tension. Try this activity to learn more!

PROJECT TIME

15 minutes

KEY CONCEPTS

Chemistry
Molecules
Surface tension

BACKGROUND

You have probably noticed that if you look at a surface outside on a rainy day or spill some water inside, the liquid tends to form droplets that stick up from a surface instead of spreading out into an even sheet. This occurs because water is made up of many tiny molecules that are all attracted to one another. Molecules in the middle of a drop of water are pulled evenly in all directions by all the nearby molecules. Those near the droplet's surface, however, are pulled mostly inward by the water molecules below them. This creates "surface tension." The surface of the water droplet is held together by the attraction between molecules.

Now, think about washing dirty dishes or clothing. There will be lots of tiny little holes and cracks that water needs to get into to wash away dirt and grime, such as the spaces between the fibers of a shirt or between a plate and bits of dried food. In order for water to flow more easily into these small spaces, you need to decrease its surface tension. You can do this by adding soap, which is a surfactant (a material that decreases the surface tension of a liquid). In this activity, you will see how soap decreases the surface tension of water by putting water droplets on top of a penny.

MATERIALS

- Penny
- Medicine dropper or eye dropper
- Glass, cup, or small bowl
- Tap water
- Dish soap
- Dish towel or paper towel
- Flat, level surface that can get wet, such as a kitchen counter
- Paper and pencil or pen (optional)

PREPARATION

- Place your penny on a flat, level surface that can get a little wet, such as a kitchen counter.
- Fill a glass, cup, or small bowl with tap water.

PROCEDURE

- Fill the medicine dropper with water.
- Now carefully add one drop of water at a time to the top of the penny. Hold the medicine dropper just above the top of the penny (not touching it) so each new drop has to fall a short distance before it merges with the drop on the penny. You can write down the number of drops you add if you like. *How many drops of water do you think will fit on top of the penny?* Watch the drop on top of the penny carefully as it grows. It should keep getting bigger and bigger until it touches the edges of the penny.
- Keep adding drops (refill your medicine dropper as necessary) one at a time. *How big does the drop on the penny get before it finally spills over the edges?*
- Once the drop spills over the penny's edge, use a towel to completely dry off the penny and surrounding surface. *How many drops of water were you able to add before the water ran over the penny's sides?*
- Mix a small amount of dish soap with your tap water.
- Now, repeat the experiment using soapy water. *Do you think you will be able to add more drops or less before the liquid spills over the sides of the penny?* Again, slowly add one drop at a time. *How big does the drop of water on top of the penny get before it breaks and flows over the edges?*

EXTRA

Try the experiment with different liquids or other things you can find in your kitchen. (Make sure you have an adult's approval to use any liquids before you handle them.) *How do different soaps and detergents like hand soap or laundry detergent compare with one another? What about other liquids like milk or juice? Which ones make the biggest (or smallest) drops? When you use small drops to cover the penny again, which liquid allows the penny to hold the most number of drops?*

EXTRA

Try using something other than a penny to collect the droplets. *What happens if you use different materials, such as the flat top of a small plastic bottle cap or a button?*

OBSERVATIONS AND RESULTS

You should find that plain tap water produces a much larger, stable drop of water on top of the penny than the soapy water does. This is because plain tap water has higher surface tension, so the surface is "stronger" and can hold together a larger drop. Adding soap lowers the water's surface tension so the drop becomes weaker and breaks apart sooner. Making water molecules stick together less is what helps soaps clean dishes and clothes more easily.

CLEANUP

Wash materials with soapy water and put them away once they are dry.

Foamy White "Snow"

YOU CAN MAKE "SNOW" ANY TIME OF THE YEAR—AND USE CHEMISTRY TO TURN IT INTO FOAMY FUN! ALL YOU NEED ARE A FEW COMMON KITCHEN INGREDIENTS AND A LITTLE IMAGINATION TO HAVE A SCIENCE SNOW DAY.

PROJECT TIME

20 minutes

KEY CONCEPTS

Chemistry
Chemical reaction
Surfactants

If you live where there is snow, it can be fun to play in it for a while—but it can get quite chilly. Or maybe you live someplace where it doesn't snow. In this fun activity, you can experience something similar to snow anytime, in the comfort of your own home. You will mix together some common kitchen supplies to make a sculpted object, and then, whenever you decide, you can let your snow creation "melt" away and turn into a white surface. Curious how kitchen chemistry can look like snow? Try this activity and find out!

BACKGROUND

Just a few common and safe household chemicals—baking soda, vinegar, and dishwashing soap—can lead to an interesting project that examines two areas of chemistry.

Vinegar (or diluted acetic acid) combined with baking soda (also called sodium bicarbonate) yields a cascade of two chemical reactions. The end products are carbon dioxide (a gas) and water, in which two chemical products (sodium ions and acetate ions) are dissolved. As you observe this reaction, you see the carbon dioxide gas as bubbles rising to the surface of the liquid and hear a fizzing noise.

Soaps cover a different aspect of chemistry. They contain surfactants, or surface-active agents. These agents reduce the surface tension of the liquid. As such, they assist in the creation of foam. Foam is pockets of air or gas trapped in a thin layer of liquid. Foam does not form on pure water, as water likes to stick together. Air bubbles rising to the water surface push through and escape. Adding surfactants to the water allows the soapy liquid to spread out and trap the air bubbles, so a layer of foam forms.

Usually, air gets trapped when we vigorously move the soapy water. Would the gas produced when vinegar reacts with baking soda get trapped and create foam? Do this activity to find out. And see what you can do with the results!

MATERIALS

- Two large plastic containers
- Measuring cup
- Baking soda
- Measuring spoons: a teaspoon and a tablespoon
- Sticky note
- Pen or pencil
- Dishwashing soap
- Vinegar
- Small waterproof items to decorate your creation (optional)

PREPARATION

- Choose a workspace that can get dirty and wet.
- Write "Soap" on a sticky note and stick it to one of the plastic containers.

PROCEDURE

- Scoop 1 cup of baking soda into both plastic containers.
- Add 3 tablespoons (44 ml) of water to the container without the sticky note and mix the ingredients together to make a dough. It should feel like play dough. You can add water, small amounts at a time, if the dough is too crumbly, until you reach the right consistency.
- Have fun molding a snowman, a polar bear, or any other critter. Feel free to decorate your creation with waterproof objects. Keep your creation in that container.
- Now, take the plastic container with the sticky note reading "Soap," and make a second batch of dough using a slightly altered recipe. Add 1 teaspoon (5 ml) of dishwashing soap to the baking soda, followed by 3 tablespoons (44 ml) of water. *Do you notice the difference in the recipes?*
- Mix the ingredients together to make a dough. You can add water, small amounts at a time, until the dough molds well. *Does this dough feel similar or different compared to the other one?*
- Have fun molding a second critter; it could be very similar or quite different from your first creation but try to make something of similar height and size. Feel free to decorate this creation with waterproof items. Keep your creation in that container.
- Admire your creations. In the next step, you will melt them in a fun and exciting way: you will pour vinegar over your creations (but don't do it yet)! *Can you predict what will happen?* Remember that the main ingredient of your dough is baking soda. *Did you ever mix baking soda with vinegar before? What happened? Would something similar happen now?*
- Fill your measuring cup with vinegar and pour all of it at once over your first creation. *What happens? What do you see and hear? Is it as you expected, or is it different?*

- Fill your measuring cup again with vinegar and pour all the vinegar at once over the other creation. *What happens this time? What do you see and hear?*

- *What is similar and what is different between the two containers? Do you remember what was different between the two recipes? Would that explain the difference between the observations?*

- Most likely, your critter is partially destroyed and only partially standing. Try something different this time; find out what happens when you pour water over your critters. *What do you think will happen?*

- Rinse your measuring cup and fill it with water. Pour all of it at once over whatever is left of your first creation. *What happens? Is it as exciting as pouring vinegar over it?*

- Fill the measuring cup again with water. *Can you predict what will happen if you pour it, all at once, over whatever is left of your second creation?* Go ahead. *Was your prediction right?*

- If your creations are still partially standing, keep trying out different actions. *What happens if you pour slowly?* Take some dough in your hand and pour vinegar or water over it as you hold it over the container. *How does it feel?*

EXTRA

Repeat the activity using different liquids to pour over the creations, such as lemon juice, soda, or milk. *Before you pour, what do you think will happen—and why?*

OBSERVATIONS AND RESULTS

Did you see both creations fizzing as soon as vinegar touched them? When vinegar comes in contact with baking soda, a cascade of two chemical reactions takes place. The result is carbon dioxide gas—a product of the reactions—bubbling up in a water solution. The bubbles create the fizz and sizzling sound. Whatever was left of your first creation (the one without the soap) was probably standing in a pool of a bubbling watery solution, where bubbles burst open as they reached the liquid surface. This was probably different for your other container. The leftover of the second creation—which had dishwashing soap mixed into the dough—was probably surrounded by a layer of white foam. Detergents have surfactants in them. These allow soapy solutions to spread out. The bubbles created in the chemical reaction still rose to the surface, but now, the soapy solution spread out and trapped the bubbles forming a foam.

As water does not react with baking soda, pouring water over your creations probably just washed away some dough without any sizzling, fizzing, or foaming.

Other acids, such as those in lemon juice, can also be used to make the dough fizz. Neutral solutions (such as water) or base solutions (such as milk) will not cause a chemical reaction. Neither creation will start fizzing when these are poured over them.

CLEANUP

Pour your liquid solutions down the drain. Any solid dough can be thrown in the trash.

React Fast

How Size Determines Rate

BUBBLE UP WITH THIS FUN TEST OF REACTION TIMES. SEE WHICH SIZE MAKES THE BIGGEST FIZZ!

PROJECT TIME

15 minutes

KEY CONCEPTS

Chemistry
Physics
Reaction
Surface area

Did you know that flour can explode? Luckily, this does not happen spontaneously on your kitchen counter, but only if the conditions are right. You need a very fine powder of flour to make an explosion happen. In fact, any solid flammable material that is dispersed in the air as a dust cloud will explode if it comes into contact with flame (a reason extreme caution must be used where there is a large amount of grain dust, such as in storage facilities). Why is that? It has to do with the particle size of the solid material, which determines how rapidly a chemical reaction takes place. In this activity, you can try this for yourself—skipping the explosion and creating a big fizz instead!

BACKGROUND

Some chemical reactions happen very fast (think vinegar and baking soda), whereas others take a very long time (such as rust forming on metal). In chemical reactions that include a solid as one of the reactants, you can actually change the reaction rate by varying the size of the solid that reacts with the liquid or the gas. How does this work? For a chemical reaction to happen, the molecules or atoms of the reactants need to collide with each other. This can only happen at the surface of the solid, as all the molecules trapped within the body of the solid cannot react until they meet the molecules of the other reactant. However, if you take the same material and break it into smaller pieces, there is much more surface area exposed that can interact with the other components—allowing the chemical reaction to occur much more quickly.

Take a flammable material (such as flour) as an example: If you heat up a lump of flour, it will not burn at all or will start burning away slowly in a controlled manner because there is only a limited surface area available that can react with the oxygen in the air. However, the same flammable material dispersed into the air as a fine powder would allow for much greater surface area exposed to the air, allowing for a rapid explosion if ignited. Increasing the surface area of a reactant does not only increase the quantity of material available to react but will also increase the rate of the reaction. In this activity, you will demonstrate this effect by measuring the rate of a different kind of chemical reaction: the dissolution of sodium bicarbonate from an effervescent antacid tablet in water.

MATERIALS

- Effervescent antacid tablets (at least four)
- Sheet of paper
- Spoon
- Four clear 12-ounce (or larger) drinking glasses
- Measuring cup
- Teaspoon
- Tap water
- Stopwatch
- Pen

PREPARATION

- Take four antacid tablets out of their packages. Take care not to break them, as the tablets are very brittle.

- Put one whole antacid tablet aside for now.
- Take the second tablet and break it into half and set both halves aside.
- Take the third tablet and break it evenly into quarters. First, break it into halves and then break the halves again into two parts. Set all four pieces aside.
- Take the fourth tablet and grind it into a powder. To do this, put the tablet to be ground inside a clean, folded piece of paper. Place the folded paper on a solid surface and use a spoon to carefully crush the tablet into a powder. Keep the powder folded into the paper and set it aside.
- Use a measuring cup to add about 1 cup (237 ml) of tap water to each of the four glasses. The temperature of the water should be the same in each.

PROCEDURE

- Take one of the glasses with tap water and the whole antacid tablet and put them in front of you.
- Pick up the whole tablet and hold it above the water surface.
- Get your stopwatch ready.
- Drop the whole tablet into the water and at the same time start the stopwatch. *What happens once the tablet hits the water? Can you see a chemical reaction happening?*
- Stir the water gently and steadily with the teaspoon. Observe the tablet closely in the water. *What do you notice about the tablet? What reaction do you think is taking place?*
- Once all the solid material of the tablet has dissolved in the water and the chemical reaction is completed, stop the stopwatch and write down the reaction time on a sheet of paper. *How long did the reaction take? Do you think this reaction is fast or slow?*

- Get a fresh glass of water and this time take the antacid tablet that you broke in half.
- Take both pieces of the tablet and hold them above the water surface. *What do you think will change once you put the two pieces of the tablet into the water compared to the whole tablet?*
- Reset your stopwatch and get it ready.
- Drop both pieces of the tablet into the water and start the timer again. *Compared to the whole tablet, do you see the same reaction happening in the water?*
- Again, stir the water gently and observe how the two tablet pieces dissolve in the water. *Do you see more or fewer bubbles forming? Do you think this reaction will be complete faster or more slowly than with the whole tablet?*
- Once all the solid tablet material has completely disappeared and the bubbles have stopped forming, stop the stopwatch and record the reaction time. *Did the reaction time change compared to the whole tablet? Was this reaction faster or slower? Why do you think this is the case?*
- With the two remaining glasses, repeat the antacid-adding steps with the antacid tablet that you broke into four pieces and the tablet that you crushed into a powder. *Do you observe any changes in the chemical reaction happening in the water? How fast or slow are these tablets dissolving compared to the other tablets? Do you notice any correlation between the reaction time and the size of the tablet pieces?*

SCIENCE FAIR IDEA

Can you think of other chemical reactions that you could use to test how the surface area of one of the reactants affects the reaction rate with water? Think of other ingredients in your kitchen that come in various sizes and forms, such as sugar crystals, cubes, or powder. *Will the same effect be observable for these substances?*

SCIENCE FAIR IDEA

What other factors can change the rate of a chemical reaction? Repeat this activity, but only use whole antacid tablets, and this time, vary the temperature of the water in which you dissolve the tablets. *How do you think the temperature will influence the reaction rate? Will the tablet dissolve faster or more slowly in hot water compared to cold?*

OBSERVATIONS AND RESULTS

Did you find that the tablet powder dissolved much faster than the whole tablet? What you probably observed in all of your trials was some vigorous bubbling once you dropped the antacid tablet into the water. Effervescent antacid tablets are made from aspirin, citric acid, and sodium bicarbonate. When sodium bicarbonate dissolves in water, it reacts with hydrogen ions from the citric acid and forms carbon dioxide. Because carbon dioxide is a gas, it forms bubbles inside the water that you can see as foam on the surface.

The fizzing and bubbling was probably more pronounced the smaller the tablet pieces were that you dropped into the water. At the same time, you probably noticed that the whole tablet took the longest to dissolve, whereas the tablet powder dissolved really quickly. This is because with smaller tablet pieces, there is more surface area of the tablet available that can react with the water, which results in a faster disintegration of the antacid tablet, as you observed.

CLEANUP

Pour the water with the dissolved antacid tablets into a sink.

Solubility Science
How Much is Too Much?

SUPER SATURATION: WHY DO SOME SUBSTANCES DISSOLVE BETTER THAN OTHERS? TRY YOUR HAND AT SOME KITCHEN CHEMISTRY AND FIND OUT!

Have you ever added a spoon of sugar to your tea and wondered why it disappeared? Where did it go? The sugar did not actually disappear—it changed from its solid form into a dissolved form in a process called chemical dissolution. The result is a tea-sugar solution in which individual sugar molecules become uniformly distributed in the tea. But what happens if you increase the amount of sugar that you add to your tea? Does it still dissolve? In this activity, you will find out how much of a compound is too much to dissolve.

PROJECT TIME

20 minutes

KEY CONCEPTS

Chemistry
Property of matter
Solutions
Solubility

BACKGROUND

Chemistry is the study of matter and how it behaves and interacts with other kinds of matter. Everything around us is made of matter, and you can explore its properties using common chemicals around your home. The way it behaves is called a property of matter. One important property is called solubility. We think about solubility when we dissolve something in water or another liquid. If a chemical is soluble in water, then the chemical will dissolve or appear to vanish when you add it to water. If it is not soluble, or insoluble, then it will not dissolve, and you will still see it floating around in the liquid or at the bottom of the container.

When you dissolve a soluble chemical in water, you are making a solution. In a solution, the chemical you add is called the solute and the liquid that it dissolves into is called the solvent. Whether a compound is soluble or not depends on its physical and chemical properties. To be able to dissolve, the chemical has to have the capability to interact with the solvent. During the process of chemical dissolution, the bonds that hold the solute together need to be broken and new bonds between the solute and solvent have to be formed. When adding sugar to water, for example, the water (solvent) molecules are attracted to the sugar (solute) molecules. Once the attraction becomes large enough, the water is able to pull individual sugar molecules from the bulk sugar crystals into the solution. Usually the amount of energy it takes to break and form these bonds determines if a compound is soluble or not.

Generally, the amount of a chemical you can dissolve in a specific solvent is limited. At some point the solution becomes saturated. This means that if you add more of the compound, it will not dissolve anymore and will remain solid instead. This amount is dependent on molecular interactions between the solute and the solvent. In this activity, you will investigate how much of various compounds you can dissolve in water. How do you think sugar and salt compare?

MATERIALS

- Distilled water
- Measuring cup that measures milliliters
- Eight glasses or cups that each hold 8 ounces
- Four spoons
- Measuring spoon
- Scale that measures grams
- Epsom salts (150 grams)
- Table salt (50 grams)
- Table sugar (cane sugar, 250 grams)

- Baking soda (20 grams)
- Marker
- Masking tape
- Paper
- Pen
- Thermometer (optional)

PREPARATION

- Using the marker and masking tape label two cups for each compound: "table salt," "table sugar," "baking soda," and "Epsom salts."
- Into one table salt cup, measure 50 grams of salt.
- Into one table sugar cup, measure 250 grams of sugar.
- Into one baking soda cup, measure 20 grams of baking soda.
- Into one Epsom salts cup, measure 150 grams of Epsom salts.
- For each cup, weigh it and write down the mass (weight).
- Add 100 milliliters of distilled water into each cup. Use the measuring cup to make sure each cup has the same amount of water. The water should be at room temperature and the same for all cups. You can use a thermometer to verify that.

PROCEDURE

- Using both of the cups you labeled "table salt," carefully measure 1 teaspoon of table salt and add it to the 100 milliliters of distilled water.
- Stir with a clean spoon until all the salt has dissolved. *What do you notice when you add the salt to the water?*
- Keep adding 1 teaspoon of salt to the water and stirring each time, until the salt does not dissolve anymore. *What happens when the salt does not dissolve anymore?*

- Repeat these steps with both cups labeled Epsom salts. *At what point does the Epsom salts solution become saturated?*
- Repeat the steps with the baking soda. *How many teaspoons of baking soda can you dissolve in the water?*
- Repeat the steps with the sugar. *Did you add more or less sugar compared with the other compounds?*
- Put each of the cups containing the remaining solids onto the scale and write down the mass (weight) of each one. *How much of each substance did you use?*
- Subtract the measured mass from your initial mass (see Preparation) for each compound. *What does the difference in mass tell you about the solubilities of each of the compounds? Which compound is the most or least soluble in distilled water?*

SCIENCE FAIR IDEA

Does the solubility change if you use a different solvent? Repeat the test, but instead of using distilled water use rubbing alcohol, vegetable oil, or nail polish remover as solvent. *How does this change your results?*

SCIENCE FAIR IDEA

Can you find other substances or chemicals that you can dissolve in distilled water? How do their solubilities compare with the compounds you have tested?

EXTRA

Solubility of compounds is also highly dependent on the temperature of the solvent. *Do you think you can dissolve more salt or sugar in hot or cold water?* Test it to find out!

OBSERVATIONS AND RESULTS

Did all of your tested compounds dissolve in distilled water? They should have—but to different extents. Water in general is a very good solvent and is able to dissolve lots of different compounds. This is because it can interact with a lot of different molecules. You should have noticed sugar had the highest solubility of all your tested compounds (about 200 grams per 100 milliliters of water) followed by Epsom salts (about 115 grams/100 milliliters), table salt (about 35 grams/100 milliliters), and baking soda (almost 10 grams/ 100 milliliters).

This is because each of these compounds has different chemical and physical properties based on their different molecular structures. They are all made of different chemical elements and have been formed by different types of bonds. Depending on this structure, it is more or less difficult for the water molecules to break these bonds and form new ones with the solute molecules in order to dissolve them into a solution.

CLEANUP

You can dispose of each of your solutions in the sink. Keep the water running for a while afterward to flush your sink properly. Dispose of all remaining solids in the regular trash. Wash your hands with water and soap.

Put Edible Paper to the Test

MAKE SOME CRAFTY SCIENCE YOU CAN TASTE—AND LEARN A LITTLE CHEMISTRY ALONG THE WAY!

Have you ever encountered "paper" on your plate—perhaps in the form of thin paper for spring rolls, sushi, or cake decorating? Did it make you wonder how this paper relates to the paper you write on? You will find out in this activity! In addition, you can discover the recipe to make the edible paper you like best.

PROJECT TIME

60 to 90 minutes

KEY CONCEPTS

Chemistry
Materials
Fiber
Starch
Food science

BACKGROUND

Paper is made up of plant fibers pressed together into a thin, flexible, but strong mat. The fibers are from a material called cellulose, which gives plants their structural support. Cellulose is the threadlike component in grains, greens, and other vegetables or fruits. It is also found in other plants as well. Wood has a lot of cellulose; in fact, approximately 50 percent of the matter in wood is cellulose. Brown rice flour by comparison contains about 2 percent cellulose.

Cellulose has many qualities that make it an excellent material for creating paper. If it is soaked and then dried, it becomes a flexible, strong material. Evaluate an unwrinkled sheet of paper—which is essentially a mat of cellulose. How smooth is it? Can it be stretched? To test its flexibility, try rolling it into a tube. To evaluate its strength, try gently tearing a corner. Observe how much resistance the paper provides.

To make commercial paper strong, manufacturers also add starches. Starches are sugars found in plants. When heated with a liquid, starches absorb the liquid, expand, and form a network of starch particles. This makes starches handy as a gluing and stiffening agent for paper.

Cellulose and starch are both edible, so homemade paper can be too. (Of course, you wouldn't want to eat notebook or other store-bought paper because it is made with other chemicals.) Is it possible to make paper that works as paper but is also tasty? Try the activity to find out!

MATERIALS

- Brown rice flour
- Potato starch
- Water
- Small bowl
- Tablespoon
- Spoon
- Microwavable plate
- Three plates
- Three sticky notes
- Plastic wrap
- Microwave
- Sheet of paper
- Pencil or pen
- Salt (optional)
- Adult helper (optional)
- Zip-top bag to store the edible paper (optional)
- Food-color (edible ink) pen (optional)
- Cornstarch (optional)
- Wheat flour (optional)

PREPARATION

- Write "Recipe 1" on one of the three sticky notes and stick it to a plate. Repeat for "Recipe 2" and "Recipe 3."
- Divide your sheet of paper into four columns. Leave the first blank, write "Recipe 1" at the top of the second, "Recipe 2" in the next and "Recipe 3" in the last column.
- Stretch plastic wrap tightly over the microwavable plate so it does not touch the bottom of the plate.
- To the small bowl, add 2 tablespoons of rice flour and about 1.5 tablespoons (22 ml) of water, as well as, optionally, a pinch of salt—this will be Recipe 1. Mix the ingredients until you have a smooth paste. The paste should have the consistency of white school glue. Add a little additional water if the paste is too stiff.
- Pour the paste onto the plastic wrap. Help the paste spread out evenly with a spoon and/or by tilting the plate.
- Put the plate in the microwave and heat the paste on high for about 45 seconds to dry the paste. (Note that the exact duration depends on your microwave.) The sheet of paper will look dry when it is done.
- Let the sheet of paper cool. The edges might curl up a bit while cooling.
- Remove the sheet of paper from the plastic wrap. This can be tricky so take your time! (Removing the plastic wrap from the plate first might help. Turning over the plate with plastic wrap and paper is another strategy.) You could also ask an adult to help with this step.
- Lay your finished sheet on the plate with the correct recipe number.
- Now it's time to make Recipe 2. Follow the same procedure but this time mixing: 1 tablespoon of rice flour, 1 tablespoon potato starch, and about 1.5 tablespoons (22 ml) of water, as well as the optional pinch of salt if you used it in the first recipe.

- Finally, make Recipe 3 using the same procedure but mixing: no rice flour, 2 tablespoons potato starch, and about 1.5 tablespoons (22 ml) of water (as well as the pinch of salt if you have used it previously).

- If any of your sheets did not turn out well, try that recipe one more time. Note that some recipes are harder to make paper from than others, but even a small sheet will work for this activity.

- Repeat the paper making procedure for each recipe one more time so, if all goes well, you have two sheets of paper for each recipe. This time, aim to produce sheets that are as thin as you can make them. It is ok if some break in the process. A small piece to test is enough for the activity, and you will get better with practice. (If you need to take a break, store the sheets of paper in a zip-top bag so they do not dry out.)

- Create at least eight rows in your table. In the first column, label the rows as follows: "easy to make," "appearance," "smoothness," "flexibility," "strength," "stretchiness," "taste" "total."

PROCEDURE

- Now it is time to test the sheets of edible paper! As you observe differences, see if you can explain why this difference occurs. Remember that Recipe 1 has only rice flour, which contains cellulose fiber, the thread-like part you find in plant material. Recipe 2 has some rice flour, and Recipe 3 contains no rice flour. The potato starch contains no fiber but does contain starch, which can provide strength and resiliency.

- Evaluate how easy it was to make each of the sheets. *Was one recipe easier to make paper from than the others? Why do you think that was the case?* Mark an "X" in the box that corresponds to the recipe that was easiest to make (or, if it was a tie, write an "X" in the columns of all of the winners).

- Observe the sheets' appearance. *Do sheets made of one recipe look more appealing than those made from another recipe? If so, why do you think this happens?* Place an "X" in the corresponding box(es).
- Compare the smoothness of the surface of the sheets. *Is there a considerable difference? If so, why would this happen?* Place an "X" in the corresponding box(es).
- Bend your papers as if you were going to roll them. *Which recipe gives the most flexible sheets?* Place an "X" in the corresponding box(es).
- Try to create a small rip in the sheets. *Are some sheets more prone to ripping than others?* Place an "X" in the box(es) for the strongest one(s). *Why do you think this happens?*
- Try to stretch the sheets. *Are some stiffer than others? Would you like your paper to be stretchy or stiff?* Place an "X" in the corresponding box(es) for the paper that has the optimal stiffness for you.
- Have a taste. *Do sheets of one recipe taste better than sheets of the other recipes? Which one creates paper that has the most pleasant feeling in your mouth?* Place an "X" in the corresponding box(es).
- Tally the Xs for each recipe and write the totals in the "total" row. *Is one recipe a clear winner for the most criteria?*

EXTRA

Are there other criteria you would like to test? If so, write them out in your table, and test them. *Do the results change the overall winner?*

EXTRA

Can you create your own recipe that is better than the best results you obtained with recipes 1, 2, or 3?

EXTRA

Think of different uses for paper—from writing to origami to food uses. *What sorts of qualities are most important for each of those uses? Can you create a recipe that performs best for a specialized use?*

EXTRA

Use a food-color pen to write on the papers or color the paper with food coloring. *Does the "ink" bleed? Which recipe is best to write on or to make colored edible paper?*

EXTRA

Switch ingredients to others that have similar fiber content. For example, switch the potato starch for cornstarch or tapioca and/or the rice flour for wheat flour. *How does a change of one ingredient change the characteristics of the paper? Can you explain why this might happen?*

OBSERVATIONS AND RESULTS

Was Recipe 2 the overall winner? Rice flour contains insoluble fibers called cellulose. These fibers make the sheets made of pure rice flour flexible, non-stretchy, and strong, characteristics we associate with sheets of paper. It also makes for a rougher surface and feeling in the mouth.

Potato starch contains no fiber. Sheets made from pure starches are flexible and smooth but stretchy. It is also difficult to make a thin sheet with this paste. These sheets do not show the characteristics we associate with paper because they do not contain insoluble fiber.

The combination of the insoluble fibers from rice and the starches from potato starch can create the desired outcome: a smooth surface on a strong, flexible, non-stretchy sheet. The reduced number of fibers also provides a smoother feel in the mouth.

CLEANUP

You can throw away any leftover paper, then wash your materials and return them to the places where you found them.

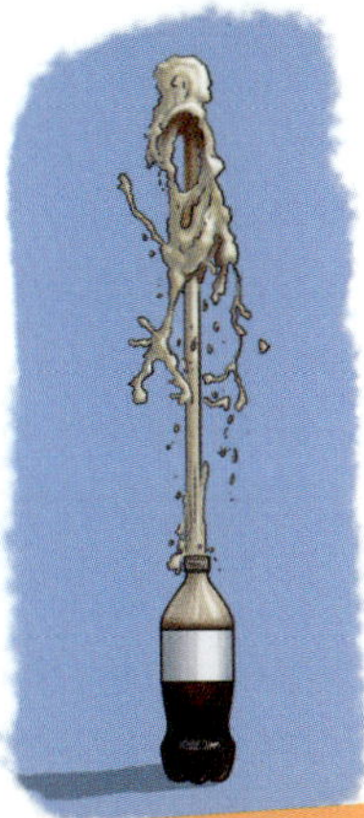

Spurting Science

Erupting Diet Coke with Mentos

USE DIET COKE AND A POPULAR CANDY TO CREATE AN EXPLOSION!

PROJECT TIME

15 minutes

KEY CONCEPTS

Chemistry
Physics
Materials science
Carbonation
Physical reactions
Explosions

Have you ever seen the Diet Coke and Mentos experiment and wondered what makes the reaction work? You might think that there is some ingredient in a Mentos candy that causes a chemical reaction with the soda pop, like the way baking soda reacts with vinegar. But the amazing eruption that takes place when Mentos are dropped into Diet Coke or other brands of diet soda pop is not a chemical reaction at all! Instead it is a physical reaction. That means that all of the pieces of the reaction are there, but that they are simply rearranged. It also means changing some factors may cause a larger or smaller physical reaction to take place.

BACKGROUND

A carbonated beverage is packed full of dissolved carbon dioxide gas, which forms bonds with water. While the soda is in the bottle, the gas is kept in solution by the bottle's pressurized conditions. When you pour some soda into a glass, some gas escapes and forms foam, but most stays trapped by the surface tension of the water. But all those gas bubbles want to escape, making it no wonder that soda makes you burp!

To create bubbles, the carbon dioxide needs to interact with itself, which means that the carbon dioxide's bonds with water in the Diet Coke must be broken. A Mentos candy can help with this. Although the candy may look smooth, if you looked at it under a microscope, you'd see tiny bumps covering its entire surface. This rough surface allows the bonds between the carbon dioxide gas and the water to more easily break, helping to create carbon dioxide bubbles and cause the classic eruption. The speed at which the Mentos falls through the soda can affect how large the eruption is, and this can be tested by comparing whole with crushed Mentos, the latter of which are less dense.

MATERIALS

- Wax paper
- Cutting board
- Knife
- One roll of Mentos (at least eight candies)
- Two index cards
- Tape
- Two 2-liter bottles of Diet Coke
- An outdoor area at least 6 feet (1.8 m) from buildings
- Eye protection (safety goggles or glasses)
- Video camera with either a tripod or a helper to take the images (optional)

PREPARATION

- Place a piece of wax paper on top of the cutting board. On the wax paper, use a knife to crush and cut four Mentos candies into many small pieces. An adult may help you cut up the candies. *What does the inside of the candies look like?*

- Make a Mentos cartridge to hold the candies for you before you drop them into the Diet Coke bottle by rolling an index card into a tube, slightly larger than the diameter of a Mentos candy. Tape the tube together on the side.
- Be sure to wear eye protection when putting the candies into the soda!
- Wear clothes that you would not mind if they get splashed with a little soda pop—this activity can get a little messy!

PROCEDURE

- Place a Diet Coke bottle in an outdoor area, at least 6 feet (1.8 m) from any buildings or anything hanging above the area, such as eaves, overhangs, or wires. Make sure that the bottle is on a level surface and stably standing straight. *Why do you think all of this is important?*
- If you want to videotape the reactions, set up the video camera so that it has in its viewfinder the bottle and a height equivalent to at least the first story of a building.
- Carefully remove the cap from the bottle and place the flat index card on top, covering the hole.
- Add four whole Mentos candies to your cartridge, put on your eye protection, and start the video camera.
- Place your full cartridge on top of the flat index card. Line up where the opening of the bottle is with the opening of your cartridge. Quickly pull out the flat index card, releasing the Mentos candies into the bottle. Then step back without tipping the bottle over or disturbing the reaction.
- *How quickly did the reaction start to happen, and how quickly did it stop? About how high did the eruption go? How much soda is left in the bottle?*
- When the bottle stops spouting, stop recording.

- Remove the spent cola bottle and place a new full bottle in the same position, again making sure that it is level and stably standing straight. As with the first bottle, remove the cap and place the flat index card on top, covering the hole.
- Add your four crushed Mentos candies to your cartridge, pouring them in from the wax paper. Put on your eye protection and start the video camera.
- Like you did before, place your full cartridge on top of the flat index card, then line up where the opening of the bottle is with the opening of your cartridge. Quickly pull out the flat index card, releasing the crushed Mentos into the bottle, then step back without tipping the bottle over or disturbing the reaction.
- *How quickly did the reaction start to happen, and how quickly did it stop? How high did the eruption appear to go? How much liquid is left in the bottle? Is it more or less than the amount that was left when you used whole candies?*
- When the bottle stops spouting, stop recording. If you videotaped the reactions, you can watch your videos now. *What do you notice from the videos?*
- *Which reaction went higher, the whole or the crushed Mentos?*

EXTRA

Find an exterior wall of a building with no windows and set a Diet Coke bottle at the base of the wall. Use a tape measure and blue painter's tape to mark off the height from the top of the bottle in foot (0.3 m) increments. Then repeat this activity three times, with the bottle in front of the tape-marked wall, videotaping it each time. When you review the recordings, use slow motion, and pause the recording when the spout is at its maximum height. Using the tape marks in the background, estimate the height of the spout. Calculate the average height of the fountains for the whole and for the crushed Mentos. *What is the difference in height of the eruptions?*

SCIENCE FAIR IDEA

What other factors affect the size of the Mentos and Diet Coke eruption? You can try testing different kinds of carbonated beverages, different kinds of candies with different shapes and textures, or using other things to start the reaction, like rock salt, pennies, or dice. *Which beverages, candies, or other things cause the largest and smallest fountains? Why do you think this is?*

EXTRA

Do this activity again but instead of testing whole Mentos versus crushed, compare warm versus cold Diet Coke. *Does temperature affect the eruption height?*

OBSERVATIONS AND RESULTS

Was the eruption higher when whole Mentos candies were used compared with crushed candies? Was less Diet Coke left in the bottle after the reaction with the whole candies compared with the crushed ones?

In the Diet Coke bottle, the Mentos candy provides a rough surface that allows the bonds between the carbon dioxide gas and water to break more easily, helping to create carbon dioxide bubbles. As the Mentos candy sinks in the bottle, the candy causes the production of more and more carbon dioxide bubbles, and the rising bubbles react with carbon dioxide that is still dissolved in the soda to cause more carbon dioxide to be freed and create even more bubbles, resulting in the eruption. Because Mentos candies are rather dense, they sink rapidly through the liquid, causing a fast, large eruption. The crushed Mentos candies, however, are not as dense as the whole ones, which causes them to sink more slowly, creating a relatively small soda fountain, which should also leave more liquid in the bottle than the larger eruption with whole Mentos candies did.

CLEANUP

Hose off any part of a building or sidewalk that was splashed with Diet Coke. If you try this project with regular Coke, the eruption should still happen, but its sugary content may make cleaning more difficult.

THE SCIENTIFIC METHOD

The scientific method helps scientists—and students—gather facts to prove whether an idea is true. Using this method, scientists come up with ideas and then test those ideas by observing facts and drawing conclusions. You can use the scientific method to develop and test your own ideas!

Question: What do you want to learn? What problem needs to be solved? Be as specific as possible.
Research: Learn more about your topic and refine your question.
Hypothesis: Form an educated guess about what you think will answer your question. This allows you to make a prediction you can test.
Experiment: Create a test to learn if your hypothesis is correct. Limit the number of variables, or elements of the experiment that could change.
Analysis: Record your observations about the progress and results of your experiment. Then analyze your data to understand what it means.
Conclusion: Review all your data. Did the results of the experiment match the prediction? If so, your hypothesis was correct. If not, your hypothesis may need to be changed.

GLOSSARY

absorbent: Able to soak up liquid easily.
abundant: Plentiful.
cascade: To occur in stages so that each stage comes from or acts upon the outcome of the preceding stage.
consistency: Having to do with the thickness of a liquid.
dense: In chemistry, having a high mass compared with things of the same size.
durability: The ability to withstand wear or pressure.
flammable: Able to be set on fire.
ignite: To set on fire.
insoluble: Not able to be dissolved in a liquid.
molecule: A group of atoms bonded together. The smallest unit of a chemical compound.
resiliency: Having to do with how strong or tough something is.
saturated: Having to do with a solution that cannot absorb or dissolve any more of a solute at a certain temperature and pressure.
spontaneously: Quickly and without warning.
stable: Not changing.
vigorous: Carried out forcefully and energetically.

ADDITIONAL RESOURCES

Books

Biberdorf, Kate. *Kate the Chemist: The Awesome Book of Edible Experiments for Kids.* New York, NY: Philomel Books, 2021.

Dingle, Adrian. *Awesome Chemistry Experiments for Kids: 40 STEAM Science Projects and Why They Work.* Emeryville, CA: Rockridge Press, 2021.

Huggins-Cooper, Lynn. *Chemistry for Curious Kids: An Illustrated Introduction to Atoms, Elements, Chemical Reactions, and More!* London, England: Arcturus, 2021.

Websites

Discovery Education
sciencefaircentral.com

Exploratorium
www.exploratorium.edu/explore/activities

Science Buddies
www.sciencebuddies.org/science-fair-projects/project-ideas/list

Science Fun for Everyone!
www.sciencefun.org/?s=science+fair

Videos

Chemistry: Easy Chemistry Experiments You Can Do At-Home | Science Trek
https://wtvi.pbslearningmedia.org/resource/easy-chemistry-experiments-at-home-media-gallery/science-trek/

Sponge Candy: Chemistry | Compact Science
https://wtvi.pbslearningmedia.org/resource/fe3c3f0f-fa71-4f02-aeea-de2693f2c564/sponge-candy-chemistry-video-compact-science/

Publisher's note to educators and parents: Our editors have carefully reviewed these websites to ensure that they are suitable for students. Many websites change frequently, however, and we cannot guarantee that a site's future contents will continue to meet our high standards of quality and educational value. Be advised that students should be closely supervised whenever they access the internet.

INDEX